Garden
SPELLS

D1631604

Garden
SPELLS

The MAGIC of HERBS,
TREES and FLOWERS

CLAIRE NAHMAD

Illustrations by CAMILLA CHARNOCK

PAVILION

GARDEN SPELLS *is dedicated to*
my mother, MAY.

ॐ

First published in Great Britain in 1994 by
PAVILION BOOKS LIMITED
26 Upper Ground, London SE1 9PD

Designed by Nigel Partridge

A CIP catalogue record for this book is available from the
British Library

ISBN 1-85793-302-8

Printed and bound in Italy by Graphicom

2   4   6   8   10   9   7   5   3

This book may be ordered by post direct from the publisher.
Please contact the Marketing Department. But try your
bookshop first.

Contents

$\mathcal{I}$NTRODUCTION

$\mathcal{A}$ garden is a holy place. From the concept of the Garden of Eden as humanity's mystic point of origin to the idea of paradise as a garden which represents the realm of final homecoming to which we strive and aspire, gardens seem to be enshrined in our consciousness as the alpha and omega of spiritual experience.

It was in a garden that Christ underwent his sleepless night of agony before his sacrifice the following day, and in a garden too that Mary Magdalene met the arisen Christ, mistaking him at first for 'the gardener'. In many religions, the idea of the garden is celebrated with reverence; and in the folklore of ancient Britain, magical associations are given to every tree, herb, bush and flower of the wayside and the garden. In its wider sense the garden embraces the countryside and – according to the wisdom of the philosophers, alchemists, hermits and wisewomen of the past – the entire earth.

From such a macrocosm, we come to the microcosmic gardens of our own making, a custom that began with the dawn of our civilization when 'gardens of Adonis' were set out in containers on rooftops in ancient Greece and bloomed in profusions of roses. Even then, gardens were regarded as a sanctuary, some peaceful, sweet-flowering green arbour where people could retire to think and to give ear to the wisdom and inspiration inherent in nature. The Greek philosopher Epicurus taught his pupils in a garden, and poets, painters and musicians throughout the ages have found their muses therein.

In ages past, intellectual accomplishments and the

opportunity for self-expression through the arts were largely denied to women. The craft of garden-making, however, was not. The garden was often their special domain and became the embodiment of their inspiration – both practical and poetic. In addition to the 'silver bells, cockle shells, and pretty maids all in a row' of the flower garden, there was nourishment from the kitchen garden, and healing and restorative plants grown in the medicinal herb garden.

The ideas of magic and enchantments, of the effect of the moon and stars on the tides of growth and decline in nature, and of fairies, elves and gnomes who were mysteriously associated with the subtle creative forces of the earth, were never far removed from the domestic garden, where all this mystery could be seen taking place. Strange magical beings were encountered: tiny flower fairies, little old men and women who seemed to have the garden under their care, tall and beautiful elf men and women who lived in trees and who could create music as fairy pipers – angelic spirits of the garden who seemed to endow the air with grace and beauty and colour – and fabulous beasts which might appear within its precincts to warn, protect or bring a supernatural message. All have been reported as living constituents of the folklore of the garden, encountered by simple country people not given to flights of the imagination.

As our understanding of the world increases, we will hopefully come to realize more fully that there is real wisdom and knowledge to be gained from the treasures of our folklore. Already we are aware that life is composed of many subtle vibrations and emanations, of which we have had little or no previous scientific knowledge, which have been preserved in the vision of wisewomen and esoteric lore. It is in the light of this awareness that *Garden Spells* is offered. It is a Victorian wisewoman's guide to the marvels, mystery and magic associated with the garden, from its trees, flowers and herbs through its animals, birds and insects and its spirits and fairy folk, to the rocks and stones that can be found in its soil. Here are the charms, spells, invocations and runes of wise-women's lore, accompanied by instructions for blessing the garden, foretelling the weather and developing a rapport with flowers, trees, fairies and animals. There are instructions for creating a herb garden in the Elizabethan mode, and tips for ridding the garden of pests and attracting butterflies.

There is true fascination in discovering the magic of these charms and rituals. For those who dislike the harsh chemical effects of modern methods, *Garden Spells* offers a new philosophy based on love and respect for all life. Its inspiration comes from centuries of homespun wisdom and traditional lore which reveal how every garden can be given a special atmosphere if its owner's heart is put into creating and tending it.

CHAPTER ONE

THE MAGICAL GARDEN

'Fairy folks are in old oaks'
(country saying)

*W*hen we create a garden, we create a rapport with nature which bestows a sense of sanctuary and well-being. To learn the magical significance of plants and flowers is to imbibe something of their ancient fragrance of mystery and benign sorcery, rooted in the breath of that great spirit which is the Earth herself, goddess of wisdom and instructress of wisewomen.

❦

THE ROWAN

If a rowan tree should take root in your garden, then your home and all who dwell therein are blessed, for the garden is under the special protection of the fairies. The rowan is a 'witch' or 'wicken' tree, which means that it is the tree of the Goddess. It is said that the wicken tree thrives upon land made sacred by ancient stone circles and forgotten Druidic rites. Should you happen upon a flourishing rowan which is most bountifully hung with cluster upon cluster of delicate red berries, then you may be sure that some saintly soul lies buried close by.

For the Garden – Boot and Shoe Rowan Berry Charm

Drop a palmful of rowan berries, gathered upon Rowan-tree Day (13 May), into any number of old boots and shoes you wish to discard – only they must be of leather. Take them and bury them deep, choosing a spot where your bedding-plants best flourish, and say this little charm over them as you work;

> *'Rowan-fruit, boot and shoe*
> *Bless my flowers the summer through;*
> *Fairies of the wicken tree*
> *Work this growing-charm for me.'*

Work by the light of a waxing moon, upon a Friday or a Wednesday, these being nights of Venus and Mercury and under the guardianship of their angels, Anael and Raphael. Then you may look forward to a veritable Eden of summer flowers.

$\mathscr{B}$AY

Wherever the bay flourishes, that garden and the dwelling it graces are protected from flash and flood. It is a tree of the old gods, a tree of the Lord, and its spirit is valiant. Its fragrance and its essence celebrate holy valour and human triumph. Take bay leaves in your food, or make of them a tisane (one teaspoonful of the herb to a cup of boiling water), for they have properties which heal and restore.

$\mathscr{F}$or the Garden – $\mathscr{H}$uman $\mathscr{H}$air and $\mathscr{B}$ay $\mathscr{L}$eaf $\mathscr{G}$arland $\mathscr{S}$pell

Take of bay leaves nine in number, and collect a winding of human hair from your brush until you have enough to pass around your hand nine times over. At the time of the new moon, or yet when it is full, tie both leaves and hair together into a garland, intoning all the while:

> 'Fair Selene, let this be
> A rope of charms and sorcery;
> I root my art in God's good earth
> To give it secret cunning birth;
> May it bloom like the bride at kirk –
> My soul and nature's handiwork!'

Bow to the silvery orb, and lay the garland in the hole you will have dug out earlier in the day, ready to receive tree, shrub, herb or flower. Set your plant with firm tenderness in the soil, speaking a blessing over it. Water well, unless there should be a nip of frost in the air, in which case you must wait until mid-morning to satiate its roots. Soon after the working of this growing spell, there will most likely come a flood of rain, which will be nourishing and beneficent for the new addition to your garden, so much so that it will be as if you caught the scent of your bay leaf charm in the rainwashed air.

THE OAK

The oak is a holy tree and is the lord of truth. There is a tradition that the voice of Jupiter may be heard in the rustling of its leaves, and indeed the oak's spirit is mighty and wise. Its kindly heart gives peace; its noble boughs give shelter. The Druids revered this tree, and the precious mistletoe to which it is host.

Charms

Carry an acorn in your pocket to protect yourself from storms, from losing your bearings and from evil intent. Paint a smiling face on your acorn. This might seem a pastime better suited to bairns, but it will make your charm stronger, for the Oak Man lives also in acorns.

Put a handful of oak leaves in your bath, and you will be cleansed both in body and in spirit. An oak leaf worn at your breast, touching your heart, will preserve you from all deception and the world's false glamour.

Carry three acorns about your person and you will have a charm for youthfulness, beauty and attainment in life. Tie and bind them with your own hair and bless them under the new moon and the full moon, every month of the year.

If you wish to know whether you and your present beloved will marry, take two acorns, naming them under a full moon for yourself and your lover, and drop them into a crystal bowl of well water. If they sail close to one another, as though knit by a bond, you will be sure to marry; but if they float away from one another, that is a token which speaks of severance.

THE HOLLY

The handsome holly is a lucky tree, for its affirms life, and is a symbol of undiminishing vitality. Ensure that it has a place in your garden, for its glowing green presence wards off unfriendly spirits. The Holly Man lives in the tree that bears prickly holly, and the Holly Woman dwells within that which gives forth smooth and variegated leaves. The first is lucky for men, the second for women. Do not burn holly branches unless they are well and truly dead, for this is unlucky.

For the Garden – Ale and Holly Berry Spell

An old charm to help your garden grow is to pour a quart of ale into a silver tankard upon the night of each new moon, and to drop therein nine holly berries, having blessed them and washed them in its rays. Hold it aloft, speaking this charm to the moon:

> 'Fair Selene, I drink to thee!
> May this mead a potion be!'

As soon as the rune is chanted, you must empty the tankard over your hollyhocks, your foxgloves and your torch-lilies; then you shall have fine blooms indeed.

THE BIRCH

The lovely silver birch is a Goddess tree, the symbol of summer ever-returning and the festival of the first fruits, which is Lammas-tide, the first day of August, when the goodness of Mother Earth is celebrated. May Day, Whitsuntide and Midsummer are also important days upon which it is well to wear a sprig of birch in your buttonhole. Maypoles are best fashioned from thorn and birchwood; upon the calends (the first day) of May, a little cluster of birch leaves, pinned as a brooch to the cloak or bonnet of a maid, will work a love charm so that she may choose her suitor in the May Games.

Charm

Folklore says that the birch is a tree of paradise, an emblem of the everlasting summer that prevails in the spirit worlds. It is said that the oak and the birch are husband and wife, and that wherever the birch takes root, the oak will come and grow nearby.

HAZEL

The hazel is a tree of kindly witchcraft and diverse blessings. The Druids held it to be the tree of wisdom and knowledge, poetry and fire, beauty and fecundity. Its nuts should be eaten by lovers, poets and scholars, although that goes for all, as there is something of each in every one of us. The milk of the hazelnut is especially blessed, and should be taken as a charm for health and fortune. Folklore says that forked hazel twigs can find gold, and that it is a lucky wood for water diviners.

Charms

Weave hazel sprigs into a chaplet and wear it in your hair; if you do this on May Day you will have good luck all year, and will have three wishes granted to you, besides, in order to fulfil your desires.

❧

To protect your home from fire and flash, make a small bundle of hazel twigs on palm Sunday, tying it about with your own hair. Keep it above the hearth, and it will bestow domestic bliss as well as safety.

Enchanter's Flowers

If you would create a garden with magic at its heart, where the fairies come to make sport at eventide, and which grows for your health and your happiness, then cherish a number of ancient herbs and flowers sacred to the months, the stars, the planets and the angels.

❦

Beans

Grow a crop of beans, for these celebrate the powers of the Goddess; the soul of this plant knows the secrets of the rites of life, death and rebirth. There is a wisdom in the scent of the blooms of the bean which only the spirit can hear; the myth of the plant is that it sings to wandering ghosts and guides them on their way to supernal realms; colliers will tell you that when the bean is in flower there will be deaths underground.

Charms

Take a stroll in the garden and inhale the perfume of beans in bloom as evening falls, for then you may be sure to dream prophetic dreams; but be wary, for you will touch the essence of your very soul with your night-time visions, and the truth in these visions is sometimes a burden of sorrow, what some call nightmares; to ride the truth of the soul can be a fearsome journey.

ℬRIONY

If you afford a little licence to the briony, you will have
on hand the mandrake or the womandrake plant, according to whether the climber is male or female. The briony
root is as magical as the true mandrake, for nature has
fashioned it just as curiously, as though it were a poppet-
doll for spell-making. The womandrake will cure all
manner of women's ills, and the mandrake chases away
rheumatic complaints.

ℱOXGLOVE

This is a fairy plant, and you will please the fairy folk if
you grow the tall foxglove to nod in your garden. Fairies
care for every flower and herb, but they find the foxglove especially hospitable. Some have seen the fairy
dwellers within, and in their fear have given the flower
strange folknames, for it is called Fairy Weed, Dead
Men's Bellows, Bloody Man's Fingers and Witch's Thim-
ble. There is a poison in the plant which causes drunken-
ness and frenzy, so decline foxglove tea, for it is sinister.
A few of the leaves and flowers of the foxglove, sparsely
scattered, will bring a sweet moodiness to your bathtime
which can do no harm.

VIOLET

A necklace of violets protects from deception and inebriation. Use them in your love philtres and in spells to restore health after long illness. If you dream of violets, fortune is sure to smile on you before long. If violets bloom in autumn, they speak a warning. A lovely myth tells us that violets first sprang where Orpheus laid his enchanted lute.

❧

HONEYSUCKLE

A posy of honeysuckle will bring a maid tender dreams of love and passion. If you bring honeysuckle into your home, it is said that a wedding will follow on its heels.

❧

PRIMROSE

The primrose is a symbol of birth and of prosperity. Count the number you first see, and if there are thirteen or more, you will be lucky all year. Laying hens are said to be influenced by this charm. If you see a single primrose, dance around it three times to avert the ill omen, for it foretells a bad laying year. Make a tea from the pretty flowerheads to soothe away sleeplessness and to dream sweetly; and a tisane prepared from the leaves (two chopped teaspoonsful) will bring back the sparkle to a failing memory and mind. Lay a little posy upon your doorstep, and fairies will cross your threshold as you sleep, to bless your house.

❧

$\mathcal{L}$ILY OF THE $\mathcal{V}$ALLEY

Country folk sometimes call this flower Our Lady's Tears, for it is said to have sprung up where her tears fell. Others call it Liriconfancy, for it is a maid's flower. Because Death seeks virgins on account of their purity, some say it is unlucky to plant out a bed with Liriconfancy alone. The distilled water of the flowers revivifies the system, and their perfume evokes images of the Goddess.

Ivy

Ivy will tell the fortune of the house. If it grows upon the walls of your dwelling, it gives protection from malice and misadventure; if it suddenly withers, the home will pass out of the present family's occupation. Wear a garland of ivy leaves about the head to prevent the hair from falling out after illness. Ivy leaves soaked in vinegar and wrapped around a corn will send it away; press out the juice from the leaves and let it be taken up the nostrils to ease a cold and a streaming nose.

An old lovespell runs thus: Pluck an ivy leaf and hold it against your heart, chanting three times as you walk:

> *'Ivy, ivy, I love you,*
> *In my bosom I put you.*
> *The first young man who speaks to me*
> *My future husband he shall be.'*

ROSEMARY

If rosemary roots naturally in your garden, it signifies that the woman of the house is a matriarch, powerful in mind and spirit, and her counsel and wisdom should be respected and followed; if such a woman moves away from the dwelling, the rosemary bush will often die. It is a holy and a magical plant, and its oil and fragrance are healing. Its name means 'dew of the sea' and its meaning is fidelity and remembrance. It is for the great occasions of life – weddings, funerals and births.

Charms

An old charm bids the bride and her groom to dip rosemary sprigs into their wine before they take their first sip, for then love will always flourish between them.

Rosemary tisane is a panacea, and a potion for beauty of the mind, body and spirit. To put a sprig of it and a silver sixpence under your pillow on All Hallows' Eve will ensure that you dream of your future spouse. If a maid sets a plate of flour under a rosemary bush at sunset on Midsummer Eve and goes again to retrieve it at first light upon Midsummer morning, she will find her true love's initials traced mysteriously in the flour.

YARROW

Yarrow is a witch's herb and a woman's herb, and should be offered in a little posy to the newly-married bride, for it brings the blessing of conjugal happiness. Yarrow is a talisman and a breaker of spells. Its tea is a panacea for maidens and mothers.

𝒱ERVAIN

This is a herb of mystery and enchantments. The Druids of old gathered it under the Dog Star, observing their secret rites and chanting sacred runes. It is said that vervain was used to staunch Christ's wounds, and sprang for the first time beneath the cross on Calvary so that use might be made of it. Today, old wives use it still to cure wounds and to banish many assorted ills. It is an enchanter's herb, and has mystic power over locks and bolts. It is for lovespells and love amulets. It is a lucky herb for brides.

𝒞harm

For all spells and charms, this rune must be chanted whilst gathering the herb:

> 'All-heal, thou holy herb, Vervain,
> Growing on the ground;
> Blessed is that place
> Whereon thou art found.'

𝓗EARTSEASE

This pretty flower with its painted maiden face is for broken hearts and for all those disappointed in love. Numerous ailments can arise to afflict some poor soul where previously he or she strode through life bonny and healthy, and the secret behind their malaise is a broken heart. Fable has it that Cupid brought colour to heartsease with one of his arrows, and indeed it is a flower for the heart. The soul of the plant is concerned with love and the healing of the heart. Country people sometimes call it love-in-idleness or the pansy.

The Seven Noble Trees
of Ireland

Oak

Apple

Alder

Birch

Hazel

Holly

Willow

The Druidical Trees of the Year

The Druids' trees marked out the thirteen 'moonths' of the year, according to the Druidic calendar:

Birch (or Wild Olive), Rowan, Ash, Alder, Willow, Hawthorn, Oak, Holly, Nut or Apple, Vine, Ivy, Reed, Elder or Myrtle

Herbs and Flowers of the Sun, the Moon and the Planets

SUN St John's wort, camomile, centaury, chicory, celandine, burnet, eyebright, marigold, heartsease, pimpernel, mistletoe, saffron, rosemary, sundew, viper's bugloss, angelica

❧

MOON White rose (garden and wild), chickweed, watercress, white poppy, adder's tongue, cleavers, willow, privet, purslane, loosestrife, costmary

❧

MERCURY Caraway, southernwood, mulberry, dill, hazel, fennel, maidenhair, marjoram, parsley, honeysuckle, lily of the valley

VENUS Elder, birch, daisy, blackberry, alder, burdock, cowslip, coltsfoot, meadowsweet, fennel, foxglove, marshmallow, groundsel, ground ivy, sea holly, mint, sanicle, mugwort, primrose, periwinkle, plantain, yarrow, sorrel, violet, vervain, tansy, thyme, valerian, catmint, wild rose

❧

MARS Toadflax, basil, thistle, broom, hawthorn, stonecrop, wormwood, lesser celandine, white mustard

JUPITER Milk thistle, agrimony, sage, balm, red rose, betony, hyssop, borage, houseleek, chervil, dock, chestnut, dandelion, cinquefoil, samphire

SATURN Mullein, bistort, nightshade, comfrey, hemlock, moss, knapweed, henbane, ivy, hemp

HERBS, TREES AND FLOWERS OF THE ZODIAC AND THE PLANETARY ANGELS

SAGITTARIUS *Flowers and herbs*: carnation, wallflower, clove-pink, sage
Trees: mulberry, vine, chestnut

CAPRICORN *Flowers and herbs*: nightshade, rue, snowdrop, Solomon's seal
Trees: pine, cypress, yew, spruce, holly

AQUARIUS *Flowers and herbs*: snowdrop, foxglove, mullein (torch flower), gentian, great valerian
Tree: pine

PISCES *Flowers and herbs*: heliotrope, carnation, opium poppy, violet

ARIES *Flowers and herbs*: thistle, wild rose, gorse,
nasturtium, woodbine, wake-robin
Trees: holly, thorn, chestnut

TAURUS *Flowers and herbs*: lily of the valley, violet, wild
rose, myrtle-blossom, coltsfoot
Trees: almond, apple, walnut, ash, sycamore, cherry,
myrtle

GEMINI *Flowers and herbs*: parsley, dill, snapdragon, fern, iris
Trees: elder, filbert, hazel

CANCER *Flowers and herbs*: poppy, waterlily, white rose,
watercress, honesty, moonwort, privet
Trees: willow, sycamore

LEO *Flowers and herbs*: marigold, sunflower, cowslip, heliotrope, forsythia, hops, peony
Trees: palm, laurel, pine, oak

❧

VIRGO *Flowers and herbs*: rosemary, Madonna lily, cornflower, valerian
Trees: hazel, elder

❧

LIBRA *Flowers and herbs*: violet, white rose, love-in-a-mist
Trees: almond, walnut, plum, myrtle, apple (apple-blossom falls especially under the dominion of Libra)

❧

SCORPIO *Flowers and herbs*: sweet basil, lesser celandine, purple heather, chrysanthemum
Trees: holly, blackthorn, whitethorn

❧

Angels

MICHAEL Archangel of the sun; his flower is the marigold, his day, Sunday

❧

RAPHAEL Archangel of Mercury; his/her flower is the iris, or the *fleur de lys*; his/her day is Wednesday

❧

ANAEL Archangel of Venus; her flower is the wild rose, her day, Friday

❧

GABRIEL Archangel of the moon; her flower is the lunary or moonwort (honesty); her day is Monday

SAMAEL Archangel of Mars; his flower is the woodbine, his day, Tuesday

እፉ

SACHIEL Archangel of Jupiter; his flower is the violet, his day, Thursday

እፉ

CASSIEL Archangel of Saturn; his flower is the snowdrop or the winter bell, little flower of hope and consolation; his day is Saturday

AURIEL Archangel of Uranus; her flower is the torch-flower and the gentian

እፉ

ASARIEL Archangel of Neptune; his flower is the opium poppy

እፉ

AZRAEL Archangel of Pluto; his flower is sweet basil and the blossom of the bean

Garden Lore

It is better if violets for planting are given as a gift, and if from a lover, all the better. Parsley, too, will grow more readily if a gift is made of its roots (the luckiest way is to lift a root from the herb-bed yourself). Beg a root if need be, for to eat a few parsley leaves every day

while you may is one of nature's best tonics. The wise
little bees know this, and the pretty herb will encourage
them to pasture all over your garden. Parsley, freely
grown among your other border plants as well as in the
herb garden, will help to make sure that its companions
do not become sickly and unclean with aphids.

&

It is best to 'steal' a root of vervain. The thinking herb-
grower understands that this is because the soul of the
vervain is so magical that all its life-cycle must embrace
an element of secrecy, especially its initiation into the
blessed soil of the earth. So that the spirit of righteous-
ness is not offended, it is as well to act out the theft as a
little drama, with the approbation of the landowner or
your neighbour (if you are lifting the root from a garden
nearby). This play-acting ritual of theft will do just as
well as if you actually stole it, for it is the spirit of
secrecy which you must foster.

&

When you plant myrtle, an old wise saw advises that the
woman who does so (it should always be a woman who
plants myrtle) will do well to spread her skirts over it
with dignity and 'look right proud'. Thus will the young
plant receive a benediction from the goddess.

&

Watch to see if the sun shines through the apple trees on
Christmas morning, and on Easter morning too, for that
foretells a plentiful, healthy crop and a year of smiling
prosperity for the owner of the garden or the apple
orchard. If the fruit is blessed by rain on St Peter's Day
(29 June) or St Swithin's Day (15 July), it will impart
the bloom of health to all who eat of it.

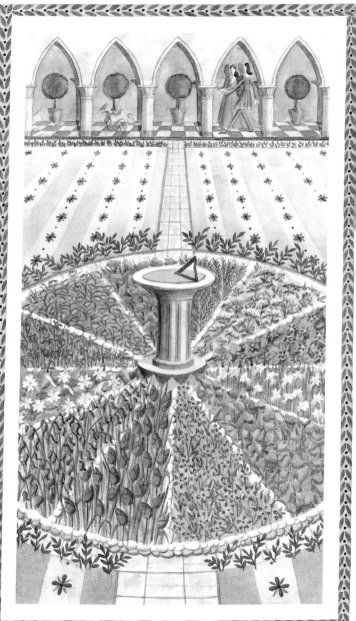

CHAPTER TWO

THE MYTHOLOGY OF FLOWERS

'Here's fetherfew, gillie flowers and bitter rue,
Come, buy my knotted marjoram, ho!'

(Streetcryer's song)

Florigraphy, 'the language of the flowers' (sometimes called 'the language of lovers'), has entranced romantics for centuries. Each flower and tree has a message, and a meaning, and is invested with a mythology which is sometimes beautiful, sometimes sinister. These strands of light and shade weave a fascinating tapestry whose rich detail rewards study, which draws us closer to the poetry in the heart of nature.

❧

THE LANGUAGE OF THE FLOWERS

Abatina Fickleness; 'I love you today, but who knows what tomorrow may bring'

Acacia (pink) Elegance; 'You possess a queen's majesty'

Acacia (yellow) Secret love; 'Let us disclose our hearts to no one'

Aconite Misanthropy; 'Your attentions please me not'

African Marigold Boorishness; 'Refinement appeals to me'

Agrimony Gratitude; 'Please accept my thanks for your token'. Agrimony was one of fifty-seven herbs in the Anglo-Saxon *Holy Salve*, which were believed to give protection from goblins, evil and poison.

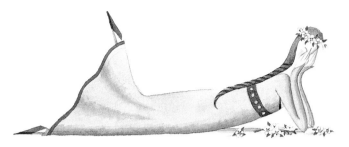

Almond (flowering) Hope; 'Your friendship is a blessing'
Alyssum Sweet virtue; 'I admire your nobility of character'
Anemone Estrangement; 'Your charms no longer touch my heart'
Angelica Inspiration; 'Your love is my guiding star'. The healing properties of this delicately lovely herb were said to have been revealed by an angel to a devout monk.
Apple Blossom Beauty and goodness; 'You are as good as you are lovely'. Apple blossom is said to fall as fragrant tears for lost love.
Arbutus Love; 'Be mine, I beg of you'
Arum (Wild) Ardour; 'My heart is aflame with passion'
Mountain Ash Prudence; 'Love should bring us wisdom, though our hearts are wild with ardour'
Ash Tree Grandeur; 'My love is lion-hearted, high as mountains, deep as the ocean'. The ash is associated with fire, lightning and clouds. In pagan mythologies it is represented as an ancestor of humankind, and Jove was said to have created the third, or brazen, race of men from ash trees; in Norse mythology, a magical ash supported the entire universe.
Aspen Tree Lamentation; 'How could you leave me thus'. The aspen is a tree of healing, and is said to shiver perpetually with horror because its wood was used to make the cross on Calvary.
Asphodel Mourning; 'Our love shall endure after death'

Balm Fun; 'I was but jesting'

Balsam Impatience; 'I can hardly live till I see you again'

Basil Animosity; 'I cannot like you'. Basil is the king of herbs and is sacred to the Hindus. An old custom advises that basil seeds be sown with 'cursing and abuse, or they will not take'.

Bay (Leaf) 'I change but in death'

(Tree) Glory; 'Your elegance and majesty dazzle me'

(Wreath) Reward of merit; 'Your determined suit has won my heart'

Beech Prosperity; 'The halcyon days of our love are at hand'

Begonia Warning; 'We are being watched'

Blackthorn Obstacles; 'Our path is beset with difficulties'. The blackthorn is a sacred tree, and is said to bloom at midnight on Old Christmas Eve, as does its 'wife', the whitethorn.

Bluebell Constancy; 'I am faithful'. The bluebell is a fairy flower, and a wish may be made on spying the first blooms of the spring.

Buckbean Repose; 'May sweet sleep attend you'

Bracken Enchantment; 'You enthrall me'. When mature bracken stems are cut crosswise close to the root, certain marks are said to appear: the initials of Jesus Christ (king of men), an eagle (king of birds) or an oak (king of trees).

Bramble Remorse; 'I was too hasty; please forgive me'. Bramble is a healing plant, and to pass through a natural arch of bramble was said to 'drive away all demons, discomfort and disease'.

Buttercup Radiance; 'What golden beauty is yours'

Camellia Loveliness; 'How radiant is your beauty!'
Camomile Fortitude; 'I admire your courage; do not despair'. The Romans anciently called camomile the 'earth-apple' because of its fragrance when walked on.
Campion Poverty; 'Though of humble station, I dare to admire you from afar'
Carnation (pink) Encouragement; 'Your charming token pleased me well'
(Red) Passion; 'I must see you soon'
(White) Pure devotion; 'A chaste love I offer you'
Celandine Reawakening; 'Let this harbinger of spring speak to you of my love'. Celandine was anciently called swallow-wort because of the old belief that the birds

made use of the flower to cure dim eyesight in their young. It is still used to improve human eyesight.

Cherry Blossom Increase; 'May our friendship wax firm and true'

Cistus Approval; 'You are acclaimed the Queen of Beauty!'

Clematis Intellectuality; 'I pay tribute to your brilliance and cleverness'

Clover (pink) Injured dignity; 'Do not trifle with my affections'

(Red) Entreaty; 'Will you be faithful to me though oceans part us?'

(White) Promise; 'I will be true to you'

(Four-leaved) Petition; 'Be mine'. Clover was sacred to the Druids; one worn in the shoe would bring the wearer's lover to them; four grains of wheat wrapped in a clover leaf gave second sight.

Cornflower Delicacy; 'Be not over-impetuous; my heart cannot be stormed'

Cowslip Winsome beauty, grace, charm, pensiveness; 'You are sweeter even than this charming spring flower'

Crocus (spring) Joy of youth; 'My heart beats with yours'

(saffron) Mirth; 'I rejoice in you'

Cyclamen Diffidence; 'I choose not to hear your protestations'. Cyclamen was considered a magical and medicinal flower, and was used in love charms, aphrodisiacs and 'love cakes'.

❧

Daffodil Rebuttal; 'I do not return your affections'

Daisy Temporization; 'I will give you my answer presently'

Dog-rose Maidenly beauty; 'You are as fair and as innocent as this pure bloom'

Dog-violet Lad's love; 'You are my first sweetheart'

Eglantine Poetry and fragrance; 'The perfume of this flower brings sweet memories of you'

Elder Zealousness; 'My efforts will remain unremitting'

Elm Dignity; 'Your queenly bearing and elegance delight me'

Enchanter's Nightshade Fascination, witchcraft; 'You have bewitched my heart with your charms'

Evening Primrose Mute devotion; 'Humbly, I adore you'

Evergreen Humility, solace in adversity; 'Thoughts of you are my only comfort'

Everlasting Flower Death of hope; 'At your wish I go away, to forget you never'

Fern Stormy passion; 'You have deprived me of your heart and left mine a wilderness'. Fern-seed was said to confer invisibility, although the spirits were so jealous of it that they tried to prevent mortals from gathering it.

(Maidenhair) Virginity; 'I am utterly yours'

Feverfew Protection; 'Let me shield you'

Forget-me-not Remembrance; 'Think of me during my absence'

Foxglove Fickleness; 'Your love is as changeable as the wind'

Fuchsia Warning; 'Take heed, your beloved is false'

Gardenia Sweetness; 'Like unto this virgin flower are you'
Geranium (pink) Doubt; 'I await your explanation'
(scarlet) Duplicity; 'I do not trust you'
(white) Indecision; 'I must ponder yet awhile'
(Wild) Steadfast devotion; 'I am your slave for ever'
Gilliflower Affection; 'You are very dear to me'
Gladiolus Pain; 'Your words have wounded me'
Golden Rod Precaution; 'Let us take care that our love remains undiscovered'
Good King Henry Goodness; 'You are the soul of kindness'
Guelder Rose Autumn love; 'He who weds me must be a man in his first youth, not one already in his dotage'

Harebell Resignation; 'I bow to your will, but hope and sigh for you still'. The harebell was considered a fairy flower, and love-divination and wishes could be made upon finding the first one, or group, of the year.
Hawthorn Hope; 'Despite your answer, I shall strive for your love'
Holly Recovery; 'I thank God you are restored to health'
Hollyhock Ambition; 'You inspire me to achieve great things'
Honesty Frankness; 'I hide nothing from you'. The 'silver pennies' of honesty, or moonwort, were believed to bring money luck if carried at the time of the new moon; slipped inside the shoe at full moon, they enabled one to see the fairies.
Honeysuckle Plighted troth; 'This is a token of my love'
Hyacinth (blue) Dedication; 'I shall devote my life to you'
(white) Admiration; 'I esteem you hightly'
Hydrangea Changeableness; 'Why are you so fickle?'

Iris (purple) Ardour; 'You have set my heart aglow'
(yellow) Sorrow; 'I mourn with you'
Ivy Tenacity; 'I desire you above all else'
(sprig with tendrils) Bonds; 'Be my bride!'

&

Jasmine Elegance; 'How dainty and elegant you are!'
Jonquil Appeal; 'Please answer soon; dare I hope you love me?'
Juniper Succour, protection; 'Please let me care for you for ever'. The juniper is a harbouring tree, said to possess an angel spirit.

&

King-cups Desire for riches 'Bestow on me the incomparable treasure of your love'

Laburnum Neglect; 'Why have you forsaken me?'
Larch Boldness; 'Only he who presses his suit with spirit shall win me'
Laurel Glory; 'Your excellence is unsurpassed'
Lavender Sad refusal; 'I can only ever be your friend'
Lilac (purple) First love; 'You are my first love'
(white) Innocence; 'A tribute to your beauty and spirituality'. The scent of lilac blooms was said to transport mortals to fairyland.
Lily (tiger) Erotic love; 'My passion burns like a firebrand'
(white) Purity; 'I kiss your fingertips'. In folklore, the lily is an emblem of innocence, purity and virginity. It is said to grow of itself upon the graves of those unjustly condemned and executed.

Lily of the Valley Maidenly modesty; 'Friendship is precious; talk to me not of love'

Loosestrife (purple) Forgiveness; 'Take this flower as a peace-offering'

(yellow) Peace; 'I am sorry; accept this flower as a token of regret'

Lupin Over-assertiveness; 'Who goes softly goes far'

Magnolia Fortitude; 'Be not discouraged; better days are coming'

Marigold Jealousy; 'Your jealousy is without foundation'. An old country name for the marigold is Summer's Bride, because it so faithfully spreads its petals at first light and closes them again as the sun sinks. The flower is a symbol of constancy and endurance in love, and was used in wedding garlands, lovers' posies and love charms.

Marjoram Maidenly innocence; 'Your passion brings blushes to my cheeks'

Meadowsweet Adornment; 'I seek a lover who is something more than merely decorative'. Those who inhaled the fragrance of meadowsweet were supposed to be given second sight and to be able to converse with the fairies. An ancient herbalist said of it: 'The smell thereof makes the heart merry and joyful and delighteth the senses.'

Mimosa Sensitivity; 'You are too brusque with my tender feelings'

Mistletoe Sweet kisses; 'I send you my kisses, as many as the stars'. Mistletoe was sacred to the Druids. Their most precious and magical plant, it could only be cut during certain phases of the moon, and then only amidst charm-chanting, with a golden sickle.

Moss Maternal love; 'How sweet is the bond between mother and child'

Myrrh Lessons hard-won; 'How bittersweet and precious
is our love!'
Myrtle Love's fragrance; 'Be mine for ever'

❧

Narcissus Self-obsession; 'You love none save yourself'
Nasturtium Artifice; 'Beauty unadorned I seek'
Nettle Coolness; 'You are cruel'. Despite its flower-
language, the stinging-nettle is a plant of healing, and old
herbals sometimes depict it as being carried to earth by
the good offices of an angel of mercy.

❧

Oak leaves Courage; 'Take heart; love will find a way'
Oak Hospitality; 'Your face and person will always be
welcome at my door'
Oats The witching soul of music; 'Your voice is sweet
music to my ears'
Orange Blossom Virginity; 'I greet you as your bride'. The
orange blossom is the bride's flower, and is said to be
unlucky for any bouquet other than the bridal garland.
Orchid Luxury; 'I shall make your life a sweet one'
Osmunda Dreams; 'Never shall I escape from your
enchantment'

44

THE MYTHOLOGY OF FLOWERS

Pansy Thoughts; 'Kind thoughts of you'
(purple) Memories, souvenirs; 'The thoughts of happy
days spent together with you are my greatest treasure'
(white) Thoughts of love; 'I cherish loving thoughts
of you'
(yellow) Remembrance; 'Oceans part us but my heart
stays with you'
Passion-flower Consecration; 'I am pledged to another'
Peony Contrition; 'I beg forgiveness for my brusqueness'.
The peony was named after Paeon, god of healing,
who used its roots to cure Hercules (who is a symbol
of humanity itself). The plant is sacred to Apollo, or
the sun.
Periwinkle First love; 'My heart was mine until we met'.
The charming blue periwinkle is known as the 'flower of
death' to the Italians, and the 'violet of sorcery' to the
French. Its associations with death and witchcraft seem to
stem from the belief that the souls of the dead inhabit
the blooms. In English folk belief, anyone uprooting a
periwinkle from a grave would be haunted by its
occupant. It was also used widely in love philtres.
Petunia Proximity; 'Always stay by my side'
Phlox Friendship, awakening interest; 'I think we could be
friends'
Pine Philosophy; 'You are the sun of my life, which
makes all things real'. The pine is the tree of the sun.
Pink Always lovely; 'No matter what the years may
bring, for me your beauty will never die'
(clove) Fragrance; 'How sweet you are!'
Polyanthus The heart's mystery; 'My spirit cleaves to
yours'
Poplar Courage; 'Let not your heart be troubled; all will
be well'
(white) Time; 'Our love is timeless'. The poplar is said
constantly to shake and whisper to itself through the

rustling of its leaves because Christ prayed beneath it when he underwent his night-time vigil of agony in the garden of Gethsemane; the tree has shivered and whispered in sympathy ever since. Poplar leaves were an ingredient of witches' flying ointments.

Poppy (red) Consolation; 'Be cheered, you still have me' *(white)* Time; 'I need time to consider'

(oriental) Silence; 'My heart aspires in silence to thee'

(pink) Sleep; 'May sweet sleep attend you, and sweetest dreams beguile your slumbers'

Primrose Dawning love; 'My heart is beginning to know you'

Primula Diffidence; 'I accept your gift with thanks'

Pyrus (Japonica) Fairies' fire; 'You are mysteriously lovely as Nature's wild heart'

Quince Temptation; 'Your charms are more than I can resist'

Reed The spirit of music; 'You are the sweet inspiration which invests my life with harmony'. Pan, god of nature, played upon the reed after the moon goddess, Diana, gave a nymph he was chasing that form for her protection.

Rose (see page 85)

Rosemary Remembrance; 'Your cherished memory will never fade from my heart'

Rue Disdain; 'Do not annoy me with your unwelcome attentions'

Rush Meekness; 'Your sweet childlike heart has utterly captivated me'

Saxifrage Humility; 'Only smile at me, and my reward
will be great'
Snapdragon Refusal; 'I cannot care for you'
Snowdrop Hope, renewal; 'I make a fresh bid for your
affection'. The snowdrop is also called February's fair
maid.
Sunflower Ostentation; 'That which glitters is not always
precious'
Sweet-pea Tenderness; 'Your memory is a lingering
fragrance'
Sweet-sultan Happiness; 'This is to wish you happiness'
Sweet-william Flirtation; 'I was only teasing you'
Sycamore Curiosity; 'I wish to know more of you'

❧

Tansy Refusal; 'Your feelings are not reciprocated'
Thorn (flowering branch) Charm and enchantment; 'I
worship you'
Thyme Homely virtues; 'I need a wife as capable as you'
Toad-flax Reluctant lips; 'Be more gentle in your wooing'
Traveller's Joy Mature love; 'Though youth has gone, my
love is strong'
Trumpet-flower Fire; 'My heart is aflame for you'
Tuberose Wounding; 'I have fluttered near the flame and
have singed my wings'

❧

Valerian Concealed merit; 'Conscious of my lowliness, I
aspire none the less to wed you'.
Verbena Enthralment; 'You have cast a spell over me'
Veronica True love; 'Nothing shall part us
Vervain Witchcraft; 'You have stolen away my soul'
Violet Modesty; 'Pure and sweet art thou'
Virginia (creeper) Everchanging; 'I am spellbound by your
"infinite variety"'

Walnut Intellect, stratagem; 'Ours would be a marriage of true minds'

Wallflower Constancy; 'Through sunshine and storm I am true to you'

Water Lily Purity; 'The light of your spirit ever shines through'

Wheat Stalk Riches; 'I offer you all I have'

Whortleberry Treason; 'You are faithless'

Willow Love forsaken; 'Be mine again'

Wistaria Need; 'I cling to thee'

Wormwood Sorrowful parting; 'Even the best of friends must say farewell'

෪

Xeranthemum Cheerfulness in adversity; 'Never say die'

෪

Yew Sorrow; 'My heart feels with yours'

෪

Zinnia Thoughts of absent friends; 'Where there is love there can be no separation'

෪

There is a lover's floral clock, which changes with the seasons. Twelve flowers are symbols of the twelve hours. The lovers' springtime floral clock can be planted out in a round flowerbed, with shells or white pebbles to divide the hours on the clock face:

1 o'clock Rosemary	7 o'clock Tulip
2 o'clock Marjoram	8 o'clock Bluebell or Harebell
3 o'clock Violet	9 o'clock Primrose
4 o'clock Jonquil	10 o'clock Clove pink
5 o'clock Sweet Pea	11 o'clock Sweet-sultan
6 o'clock Herb Robert	12 o'clock Carnation

CHARM FOR SOWING SEEDS

To be used at the time of the new moon.

'I drop these seeds in Christ's good name,
Pray Sachiel bid them rise again;
I call upon the fairy host
On Father, Son and Holy Ghost
Bless each one with hearty growth
Banish death, disease, mishap and sloth;
God bless the soil, God bless the seed,
New moon, new moon, God bless me,
God bless this house and family!'

FLOWER DREAMS

To dream of gathering flowers indicates a delightful surprise.

To dream of a basket of flowers indicates a birth or a wedding.

Dream of a wreath of flowers, expect a new love.

To dream of withered flowers is lucky,

To dream of smelling flowers means you must grasp opportunity.

To dream of gardens is a blessing for your spirit, and indicates philanthropy.

OF FAIRIES, SPELLWEAVING, AND THE ANGEL OF THE GARDEN

'The faery beam upon you,
The stars to glister on you;
A Moon of light,
In the Noon of night,
. . . And the luckier lot betide you.'

(Ben Jonson)

To accept that fairies do indeed exist, and that co-operation with them in the creation of a garden will richly benefit all that grows in it, is a part of the wisewomen's ancient doctrine.

However small and humble your plot, there is always an angel of the garden. She must be called upon, and the spirit of her gentle presence must be felt and revered by all who would make friends with the fairies. Seek her at eventide, or as the sun is rising, or yet in the first moments of daybreak, or under the moon and the stars on a fine, clear night. Speak to her in your own words; alternatively, you may use this invocation:

'Angel of the garden, still my waiting soul, so my eyes may see thy radiance, my spirit enter into thy peace, as thy wings untold to gather this place of tended growing things into thy heart; I feel the breath of the angel as a soft incense moving through the airs which play upon this garden; I feel the touch of the angel as each blade of

grass, each sprig, flower, herb and tree, each living
creature, is blessed; I hear the song of the angel as each
plant moves and dances, with a motion unseen, to the
perfect harmony of the spheres. The life of the spirit
walks in my garden as an angel, and my garden is made
holy, a place of benediction. The angel of the garden
hears my prayer, and draws near. In reverence I bow to
the angel, and give my heart and hands into the light of
her inspiration.'

When this invocation has been spoken, you must sit on
awhile so that you may experience the presence of the
angel ever more strongly and certainly. Each time you
invoke the presence of the angel, you are blessing your
garden and yourself.

There is a great 'Earth Angel', one of the four, whose
flowing robes are the very life-stuff of the natural king-
dom. Every angel, every fairy and elemental being works
under her guidance, and the angel of your garden is one
of these. The angel gives to your garden its spirit and its
peace, and directs the little people, the fairy folk, in
their work as they tend each flower and tree.

If you would know and work with the fairies, you
must seek to draw near to them. They are curious about
human folk, and are ready to love them; but they are
also afraid and angry. With our dirty, greedy ways, we
commit abomination in their world. To soothe them, we
must attune ourselves to Nature, and do things her way.
When we pour out love into the heart of the garden, and
cherish each living thing in it, both humble and magnifi-
cent, beautiful and ugly, then we may begin to see the
fairies. Peep at them from the corners of your eyes, as if

they were shy woodland creatures which you might frighten away with a direct and penetrating glance. Look deep into the lovely forms of the flowers and the trees you care for, and you will see the essence of their spirit. The colour of each plant glows radiantly, because each reflects the hue of the nature-being that brings their life-energy to them. Every differing shade of colour has a story to tell of the inner worlds. Look particularly at apple blossom in the springtime, and let your fancy weave tales of the Goddess, for the loom of the imagination brings truth to our hearts.

To open your heart to the fairies, you must nurture these feelings of wonder, reverence and love for every detail of your garden, for the airs which blow about it, the musical rain which falls gently upon it, the moon and the stars which silently look down on it, the great sun which is the source of its being, and for the clouds and the changing skies which provide it with a canopy. When you can truly feel the sweetness of this magic, you will begin to discover the fairies, for they will make themselves known to you.

These are certain charms and spells you may use to attract fairies to your garden. One ritual to ensure that you keep on good terms with them runs thus: Be sure never to uproot or otherwise kill an established plant or tree; when you remove what are known as 'weeds', do it kindly, and never when the herbs are in bloom. If you can put them to good use for food, medicine or craft-working, then the heart of Nature is appeased. If not, let them rot as compost, to be recycled according to Nature's plan. Everything that is uprooted must be treated with care and respect, and thanked for its appearance in your garden, however inconvenient that may seem. Each 'weed' in our garden can be used for the good of our health, or for the healing and restoration of the soil. Tend and talk to each plant and tree, give them courage to grow. Thus you will touch their spirit, and the spirit of the garden likewise. Do not listen to those who tell you that this behaviour seems a mite touched, for your garden will flourish like paradise, whilst theirs remains commonplace.

Walk often in your garden, for walking is a magical act and can be made sacred. Walk barefoot when the season permits. As you walk, bestow blessings and joyful reverence from your own heart upon all that you see, upon each plant, mighty and small, and upon the garden as a whole. Let each footstep convey these things into the very soil, and, as you go, you may try 'flower-breathing', which is the regular inhalation of the blooms so that their fragrance passes deep into your soul. Many have been healed in this way, especially when flower-breathing the perfume of roses. Keep your breathing easy and steady, each one slow, regular and lasting, yet not too deep. In this way, your flowers will come to know and love you, and you may stand at your open door in the early evening and ask that they send their perfume towards you. If you wait with a still soul, you will soon begin to sense their sweet fragrances on the air, surrounding you and caressing you.

If you should leave your home and garden for a while, take care to let all your flowers, shrubs and trees know of it, and reassure them of your return, or they will become despondent. On your arrival home, steal away into the garden to greet them, and all your weariness will melt away.

In this wise, observing such rituals, you will please the fairies well, and the little people will become your friends. Here is a charm to call to them:

'Fairy host, from the wild
Come and tend this plot awhile;
Come dancing from the hollow hill
To raise the power and do God's will;
Make your revels in my garden,
May this soil be fairy-trodden!
Each herb and flower, each garden tree
Set each lovely spirit free!
May all be hung with globes of light
From deepest Elfame, fair and bright.

Fairies, heed this pledge I tell
To honour you and treat you well!'

To work a fairy spell, pluck vervain and yarrow, mistletoe and rue, thyme and bay, dice each leaf and bake them into a little oaten cake, which must be sweetened with honey and three drops of rose-oil. Take it, freshly baked, at the time of the full moon, or at moonrise on Lady Day, Walpurgis Night, May Day, Midsummer's Eve, Candlemas Day or Lammas-tide, Christmas Day, Christmas Eve, Hallowe'en, All Saints' Day or Easter Day, Whitsuntide or Midsummer's Day, and set it under a tree or a bush in a little wild spot in your garden, or just beyond its boundaries. Bless the cake and say:

> *'Fairies, the work of my spirit I give thee,*
> *Be lovers true to my garden, I bid thee.'*

If the cake can be placed as you watch the moon rising, and if it be a waxing moon, that is all the better. You will know if your craft is good, for you will begin to see a new radiance steal into the blooms, and a fresh vigour vivifying all the garden. Furthermore, you will perceive fairy rings where the folk of Elfame hold their revels. It may be that on still summer nights, or yet at dawn on a spring morning, you hear a fairy piping, which is the wildest, reediest sound mortal ears ever gladdened to. You will notice their woven baths in the bushes, which are like tiny, silvery hammocks sewn from spiders' webs; these are sustenance for their own dancing, luminous selves. And you may note that your flowers and trees, of themselves, begin to form natural bowers and arbours, exquisite in their beauty and magical artistry, fit for the finest queen.

HEALING THE GARDEN

There are spiritual worlds where angels dwell, and these worlds may be entered into even whilst we walk this earth. From these bright airs of paradise, we can call upon angelic help to heal and bless the garden. The fairies too will respond to the thoughts of healing and blessing that you send forth. To heal your garden, you must still your mind and your soul and summon to yourself a golden vision of a six-pointed star, pure and brilliant, blazing in the skies and pouring down upon the garden an effulgence of light. Breathe in and breathe out this divine radiance, and you will feel it fill up your heart. This precious gift you can impart to your garden, for it will flow forth from you with every out-breath, and you may direct it deep into the soil. When you have finished your meditations, ask for a blessing from the angels of the earth, and from the fairies.

*W*EEDS

Now be certain not to clear away or clip back those wild herbs the ignorant call 'weeds' when they are in full flower, because this will cause great hurt and insult to the fairies who live at this time within the blooms. Take out those herbs you would discourage as they come up, cleanly; and if you bypass any, then let them stand and sway till just before they go to seed. Yet think twice before disposing wholly of these, for vigorous and resistant growth in your garden means that the plants have come to help you. If you have any ailment or discomfiture, then it is more than likely that these very herbs you fancy such a nuisance are those which can best help your recovery. The stars command certain herbs to grow in our gardens, for our protection and benefit; and none of us, however fine a gardener, has greater wisdom than the stars.

Take good care not to uproot an elder tree. If you must do it, then let it only be in a spell of wet down-pouring weather, in the autumn, and even then the tree must not be very large. I counsel you to go with your spade and plant it again in the wilds; and this you must say to the Elder Mother who lives inside the tree:

> *'Elder Mother, within the tree,*
> *No disrespect I mean to thee.'*

Then, as you replant the elder, you must murmur over this rune:

> *'From Christ's sacred heart*
> *Fell three drops of blood*
> *And Mary's Tears*
> *Fell into the mud.*
> *Water, Spirit, and Body of Earth*
> *Bless this tree with second birth.'*

Couch grass is good for many human ills, but to stop it rampaging sow lupins and tomatoes where its hold is most stubborn. The pretty marigold will also discourage the sprouting of weeds, and marigold tea will keep away tumours and beautify the complexion. Grow marigolds liberally everywhere, and there will be no need for harsh disposal of weeds. Foxgloves grown freely in the garden will keep weeds and disease at bay, and make the flowers and the trees happy. Grow a clump of camomile in your cabbage patch, for then your vegetables will sweeten and flourish; and if you have a sickly plant or young tree which seems to be gasping its last, take a root of camomile and let it grow with the invalid. The failing plant will rally, but then the camomile must be taken away again, to be replanted amongst the cabbages. Or, it can be used to make camomile tea, which soothes and restores the spirits, and takes away weariness.

ℱ*AIRY* ℱ*REES AND* ℱ*LOWERS*

The fairies care for all growing things, and these folk swarm over the garden in little bands. But some fairies are deeper and nobler of spirit than their fellow tiny beings, delightful though these are. The individual fairies are taller, more beautiful, mysterious and magical, of greater stature than humans sometimes. Often they will take on the guise of a little old man or a little old woman who appears to the keen and tender gardener and gives wise advice and secret knowledge of the life of the garden under that gardener's care. Sometimes, too, they will slip over the threshold into a human dwelling, if the door is left open to the garden in the cool of the evening, or at noon, or in the early morning. Then your home will feel enchanted, and you will know that a mystic visitor walks abroad, suffusing the house with a joyous magic. There are a number of flowers and trees which the fairies love particularly, and these are as follows:

Rowan, Periwinkle, Cowslip, Bramble, Cuckoo flower, Henbane, Ash, Bean, Clover, Wild rose, Cyclamen, Daisy, Oak, Valerian, Yarrow, Garden rose, Rue, Morning Glory, Thorn, Vervain, Lavender, Briony, Elder, Primrose, Birch, Thyme, Fern, Mandrake, Foxglove, Snowdrop, Apple, Rosemary, Bracken, Celandine, Hazel, Snapdragon, Aspen, Bay, Betony, Chicory, Lily, Springwort, Lilac, Nightshade, Hemp, Meadowsweet, Honesty, Honeysuckle

*P*RECIOUS *S*TONES

If you can manage to procure rough chippings of precious and semi-precious stones, then do not be afraid of burying them in your flowerbeds, for the stones of the earth, even the rocks and pebbles, are alive and quickened with spirit, and their heartbeat can be sensed by the wise and the knowing. Gods and spirits of great beauty dwell within the inner magical world of precious stones, and the place where they are buried will produce harbours of peace and goodness, where flowers will grow and absorb special vibrations from the crystals. These flowers may then be cut for the house so that they might impart their influences, strengthened by the abundant graces which precious stones enshrine. Murmur this rune as you bury the stones:

> 'Gods and fairies in these stones,
> Shine forth amidst Earth's soil and bones;
> Let your essence bless these flowers
> That their hues might wear your powers.'

*F*AIRY *D*REAMS

A number of flowers there are which it is said the fairies bring into our dreams, so that we may guess something of our future and the destiny of our souls:

Clover A happy and prosperous marriage; *May-blossom* A new lover; *Marigold* Coming wealth; *Periwinkle* A spirit watches over you; *Lilac* Luck in love; *Lily* Good luck; *Honesty* Money-luck; *Nutmeg* A change for the better; *Rose* Your heart's desire will come to you; *Myrtle-blossom* A wedding; *Apple-blossom* A birth; *Rosebuds* Many blessings; *Rosemary* Everlasting love will be yours; *Sage* You will marry; *Windflower* Your love is untrue, seek another;

Poppy A message; *Bluebell* A stormy and passionate love-affair is foretold; *Buttercup* Business success; *Carnation* A secret admirer; *Crocus* Danger in love; *Primrose* A sweet new friendship; *Snowdrop* You will have secrets to share; *Iris* A letter; *Forget-me-not* Find a new love; *Honeysuckle* Tears and smiles; *Geranium* A soon-mended quarrel; *Peony* You must grasp opportunity; *Violet* Your spouse will be younger than yourself; *Daffodil* You are neglecting a friendship — make amends.

*T*HE *Z*ODIAC AND THE *M*OON

The moon traces her path through the zodiac once each month. She is the mistress of physical form, of bodies, and her creative powers are drawn from the magical influences of the stars and planets, and from the inspiration of the particular sign of the zodiac through which she is passing. She spends two-and-a-half days (or thereabouts) in the grand star-court of each zodiacal sign.

The moon especially is queen of the ocean, of bodies of water, and her waxing and waning must be taken account of by the wise gardener. Harvest your vegetables and fruits just after the full moon, when its first light is still barely visible. Let your planting be done at eventide in a waning moon, for then water will run to the roots. So seeds during the time of the waxing moon, one day after the new moon's crescent has smiled in the skies. Prune two days after the new moon when you wish more growth; if you desire less, prune in the wane of the moon. To follow this counsel best, you must procure a good almanac. Then you will be able to note also the sign of the zodiac through which the moon is passing.

Choose air and fire signs for contemplation and meditation in the garden. During air days, concentrate most particularly on communicating with the spirits of the trees, the inner essence of the flowers, the fairies and the elves. Make joyful contact with the sylphs of the air, for they will keep you in good humour, and bring laughter to your heart. Stake climbers in air days, and sow flowers. Sow ramblers in Gemini, and flowers in Libra. In Aquarius, meditate upon the brotherhood of all life as it is expressed in the garden: human, soil, water, trees, flowers, moon, stars, stones, skies and the fiery sun are all one brotherhood. Feel the presence of fairies and angels within the glorious scheme of this mystic brotherhood. During Aquarius, if the weather is fine, sleep for a little beneath the boughs of a tree, and let the spirit of the tree speak to you in your dreams.

Fire days are for spell-weaving. Heal and bless your garden, and chant your runes. Sow plants in Aries so that their growth is fulsome; for burgeoning growth and height, sow in Sagittarius. Leo is often too hot, so spend these days designing and planning.

Water days are for wet work, so water well and plant at this time. Sow during the water signs, but never harvest in Pisces, for your fruits and vegetables will rot away quickly. Vegetables are best harvested in Aries, and fruit in Taurus. Harvest in the last quarter of the moon, when it is in decline, if the fruit is to be kept, for then the juices will be less than in the penultimate quarter. For sweet leaves and succulent fruit, sow on water days, but remember that anything sown or harvested in Pisces cannot be preserved.

Earth signs are for wholesome toil, for spadework and sowing, thinning and transplanting. Weed in Virgo, feed in Capricorn in a waning moon (watery Scorpio is also a good time to feed), and plant potatoes, corms and bulbs in a waning moon also, upon the earth days. Gather seed in the last quarter of the moon, in Taurus or Capricorn. Herbal sprays for pest control are best administered in Virgo. Any vegetable that is hot on the tongue is best harvested in Scorpio or Leo.

Earth days are also efficacious in promoting communion with the fairies. If it rains upon a water day, go out into the garden and enjoy the company of the undines, the water fairies, as they laugh and tumble in the drops. It will take the dreariness of the rain away. When you light a fire to consume garden rubbish, think of the salamanders, the fire-folk, dancing in the flames. It is good to light a fire often in the lingering summer dusk, and watch beside it whilst the stars come out.

Water Signs
Cancer
Scorpio
Pisces

Earth Signs
Taurus
Virgo
Capricorn

Fire Signs
Leo
Aries
Sagittarius

Air Signs
Aquarius
Libra
Gemini

CHAPTER FOUR

GARDEN CREATURES

'Observe which way the hedgehog builds her nest,
To front the north or south, or east or west;
For if 'tis true that common people say,
The wind will blow the quite contrary way.
She has an art which many a person lacks,
That thinks himself fit to make almanacs.'

(From *Poor Robin's Almanack*, 1733)

If you can leave a little corner of your garden to grow wild, this will please the fairies, and many little animals will visit by day and by night. Some of these may partake of your seeds and flowers, yet this is only Nature's wise plan, for moles improve the soil and birds assist pollination. If infesting insects mar your flowers, infuse horsetail and stinging nettle, and use it as a spray, for this gentle cleanser works in harmony with nature. Each insect and animal plays its part in the great scheme, and its soul has mystery. Even the humble crawling beasts bear secrets within them to which we should pay reverence.

ANTS

In some parts, ants are called Meryons, and folklore says that their kingdoms underground are fairy kingdoms. Their busy life, full of purpose, has an air of strange sanctity if trouble is taken to study it. Do not kill them, but rather ask them to leave your garden if their presence becomes a nuisance. The asking must be done three times over, in a waning moon, and must be spoken directly over the bustling ants. Say this charm, in a voice clear and firm, on three successive nights, each time repeating it thrice:

> 'Pismires, with blessings I greet thee,
> Change your abode, I sternly entreat thee!'

If the asking is done in a proper spirit, which is one of authority and respect, the little ants will soon be gone.

BEETLES

Never kill a beetle, but leave him to go about his important work in the garden. Folks say bad luck and seven days' soaking rain is the penalty for stamping on a beetle. This is because cruel behaviour angers the fairies, who can visit bad luck on us. If a black beetle crawls over your shoe, it is a warning against illness which bids you to take better care of your health. Many nocturnal flying beetles predict fine weather.

SPIDERS

Spiders are lucky, so, as the wise-saw bids, 'If you wish to live and thrive, let the spider run alive.' Spiders are sensitive to human feelings, they can be spoken to and will respond. They are friends of the garden. It is said that during the biblical Flight into Egypt, the Holy Family took shelter in a cave. A spider worked industriously until it had woven a web so thick across the entrance that a dove came and laid an egg in its silver strands. When Herod's men, in pursuit, reached the cave, they passed it by, thinking that no one could be within. Many folktales tell of such little creatures helping human beings in their time of need. Such help always seems to come if the one in need has shown kindness and mercy to living creatures. There is much truth enjewelling these old stories, if we could but listen with an open heart.

The spider bears within itself the gifts of healing and wisdom. It spins a web of life and sits at its centre, like the Creator. It is a sacred symbol of the Goddess, spinning her threads of human destiny. In the house, it tells of prosperity and happiness; if it drops down upon you, that is money-luck; if it runs over your clothes, you will soon have beautiful new garments. To catch one and put it in your pocket for a moment (without harming it) is a charm to make your pocket jingle with silver. Spiders are also prophets of the weather, and when the are industrious in their web-building that means fine weather; when stormy weather threatens, they do not extend their webs, but make haste to strengthen them. Long, fine, flowing threads mean a spell of sunny weather; short threads foretell wet and boisterous conditions, prevailing for a day or two. When they take refuge in nooks and crannies, having broken down their webs, expect persistent rain. If they alter and rebuild in the evening, the night will be calm and clear; if they begin a web in the morning, the day is sure to be fine.

❧

Snails

Snails are symbols of unhurried patience, and they will eat the moss which gathers on the boles of fruit trees, thus helping the gardener. A springtime charm is to place one among the cool white ashes of the previous night's fire, which must be spread on the hearth. The snail will trace the initials of your true love.

❧

Ladybirds

The ladybird is a symbol of fire and the sun. If a ladybird alights on your clothing, you will soon wear your

wedding-dress. If it settles upon your flesh or hair, good fortune is prophesied: you will have as many good months as there are spots. An old charm for marriage divination runs thus:

'Bishop, Bishop Barnabee,
Tell me when my wedding be;
If it be tomorrow day,
Take your wings and fly away.'

A ladybird with seven spots on its back is a fairy's pet, and you may make three wishes before it flies away.

❧

GRASSHOPPERS

These lively little creatures foretell travel and good news.

❧

WASPS

A wasp sting warns you to be on your guard against jealousy, deception, danger, and the bearing of grudges.

❧

BEES

Bees are said to come from paradise, and to have a special connection with the planet Venus. They are traditionally revered as divine messengers, foretellers of the future, wise guardians of the secrets of nature, and little winged servants of God who hummed anthems of praise. This they do with special fervour upon Christmas Eve, humming the hundredth psalm at midnight to honour the birth of Christ. They are the little priests of the garden, and they cannot bear blasphemy or swearing. They love purity, and tradition says that a virgin can walk unharmed through a swarm of bees. If you have a hive, you must tell the bees the family news, or they will leave. They cannot remain where there is anger or hatred in a household. When a member of the family dies, the iron door-key must be struck three times upon the hive, and they must be told, by name, of the person who has died. They must then be allowed a period of mourning.

It is lucky for a bee to come into the house. A bee alighting on the hand predicts the arrival of money, and one settling on the head means that you are destined to rise to glory. If a bee flies around a babe asleep in its cradle, that child will be blessed with healing and prophetic powers, and will have a happy life.

ℬATS

When bats rise heavenwards in their flight and then drop swiftly again, it is said to denote that the witching hour has come, and all benign charms and spells will work especially well at such a time. Evil spells will never work well, for their desserts will be visited on their creator.

It is a sign of extraordinary good luck if a bat touches you in its flight. A charm to court the good luck runs:

> *'Airy mouse, airy mouse, fly over my head,*
> *And you shall have a crust of bread,*
> *And when I brew and when I bake*
> *You shall have some wedding cake.'*

ℋEDGEHOGS

A hedgehog makes a sweet-natured friend, and a little cream put out for it on a dish at night will ensure its nightly visits (milk upsets their digestive systems). Country folk call it the Urchin. It is a goodly little beast, and very affectionate. It knows when storms are brewing, and can tell the direction of the wind, building its nest in accordance with its wisdom. It is an an animal of good omen. To overtake the Urchin is lucky, while to meet it going in the opposite direction is even more fortunate.

※

ℳOLES

It is lucky to find a molehill, though unlucky for your lawn. Sink an upright glass bottle into it, and the sounds the wind makes in the glass chamber will encourage the mole to set up home elsewhere. Use the soft fresh earth he has turned up for potting.

※

ℱROGS

To find a frog in your garden is lucky. Look for the frog-stone, which is a yellow stone shaped something like a frog. These are lucky, and should be kept, for they contain secrets which the spirit can hear when the stone is held in the hand.

*T*OADS

If a toad should take residence beneath a stone, or amongst some old tree roots, honour his jewel-eyed presence in your garden. Do not call him names or drive him away. He is a custodian and guardian and within him lives the spirit of the garden. He will keep away all pestilence and attract the fairies, and he will ward off distemper within the home. It was said in olden days that a precious stone lived within his head, with a guardian angel alive inside it. If you make him your friend (and this can be done) you will be able to look into his eyes; and you will see that indeed they are like a magical composite of moonstone, topaz and soft brown agate, and that in them seems to dwell the spirit of wisdom.

❧

*M*AGICAL *B*IRDS

CROW Bird of wisdom and the secrets of life and death. He serves the Underworld and is a king in his own right.

ROOK If you have a rookery, these are birds of fortune for you; they must be told the new of the household, and must never be driven away.

BLACKBIRD Of the crow family, but a magical sorceress because she has learned to sing sweetly; some call her a witch. She is a luck-bringer, and can help to enchant a lover if her arts are called upon. Her song, like all birdsong, is hallowed in the morning and in the evening.

CUCKOO A rain-bird, a representative of the Goddess and the spirit of spring; at the first call of the cuckoo, make a wish. If you hear her whilst standing on the grass, the coming year will make you rich.

DOVE A happy-omened bird, signifying love, joy and wedded felicity. The dove is the bird of Venus and is very lucky for lovers.

OWL The bird of enchantment and witchcraft. She can tell your future if you put your questions to her. One hoot is no, two mean yes, and more mean you must put your question again another time. No significant pause should occur between the calls when you are divining; if it does, the answer is just one call, which is no.

SPARROW If a sparrow flies into your house, it comes to warn of an illness. It is the spirit of merry mischief and is lucky for lovers.

FINCHES All the finch family are birds of good omen, and there is a magical belief that if you tell them your

75

troubles, they will fly away with them, and thereafter your life will be healed.

TITS The tit family are also happily-omened, and if they fly much about your dwelling in the winter, your home and garden will be protected from winter's ravages, and there will always be fuel for the fire.

ROBIN The robin is a holy bird, and one which loves gardeners, for they remind him of paradise, which is his true home. He brings good luck, Christmas cheer, and a prosperous New Year.

MAGPIE A bird of duality, of good and evil. To see one is bad; to avert the omen, call out 'Raven, seek thy brother.' To see two is very fortunate. The bird gives augury according to the famous children's rhyme, *One for Sorrow, Two for Joy.*

SWALLOW One of the most fortunate of birds, the symbol of spring and regeneration, birth and awakening. To see a swallow in the early days of spring, before they are numerous, is very lucky. If swallows nest in the eaves, success, happiness and good fortune are assured for all the inmates, and your house and garden trees are protected from storms.

SWIFT A token of the summer, and a symbol of the human soul; it is the bird of tears which can be both sorrowful and joyful.

HOUSE-MARTIN This bird brings protection from thunder; it is also a bringer of domestic felicity, fecundity, and romance to those who are single; it is the bird of house and hearth and homely virtues.

WAGTAIL The dapper little wagtail is the bird of peace and content, and to see him hopping about in your garden is a lucky sign.

WREN The wren, bird of life and hope, is a very fortunate creature, and a welcome sight in the garden, promising fulfilment.

THE MAGIC RING OF SOLOMON

King Solomon was said to have a ring which enabled him
to speak to the birds of the skies and the creatures of the
earth. This 'magic ring' is a ring of faith. It is composed
of spiritual light. You can create it as a magical seal for
yourself if you are kind and gentle towards our animal
and insect brethren.

Sit at your altar at ten o'clock in the evening, which is
Gabriel's hour, and light a white, a green and a brown
candle. It must be the night of the new moon, and you
must bow to the angel of the moon (Gabriel) before pro-
ceeding. If you can manage to have a sapphire close at
hand (the 'seal of secrets') that is all the better. Speak
this charm:

'King Solomon, King Solomon, I seek your magic ring;
So I may talk with the beasts, and hear what the birds sing.'

Now call upon the four Angels Chasan, Arel, Phorlakh
and Talishad to help you in your creative endeavours.
Only remember this is a working which must flow forth
in purity from the heart. Repeat the ceremony once again
at full moon, and a third time upon the night of the next
new moon, and the task is done.

Remember never to use hurtful means to discourage animals in the garden. Call always upon the angel of the
garden to help you to speak to them and bid them leave.
In this way, you will begin to see the fairies, and will
wear Solomon's charmed ring.

OF GROTTOES AND GARDEN NOOKS

'God Almightie first planted a Garden. And indeed,
it is the Purest of Humane pleasures. It is the Greatest
Refreshment to the Spirits of Man; without which
Buildings and Pallaces are but Grosse Handy-works . . .'

(*Essaye XLVI*, 1625, Francis Bacon)

When planning your garden, it is well to know which plants thrive happily together, and which flowers disdain each other.

Daffodils do not like tulips, who in their turn dislike lilac. Lilies-of-the-valley must nod in a bed of their own, for other flowers do not care for them. Poppies are loth to allow other flowers to flourish – do not let them grow too near their companions. Roses dislike carnations and mignonettes, although they will tolerate marigolds, parsley and lupins. Do not place a vine near the cabbage-patch (camomile is helpful there, though) or close to the bay-tree, although it will tolerate elms, poplars and poppies. Oak dislikes walnut and the olive, but the olive loves myrtle, maple and the fig tree.

The birch and fir love one another's company, and the oak will grow near to a birch wood, though not too close. The blackthorn and the whitethorn share a spiritual affinity, but cannot bear to grow close to one another. Lavender dries up cucumber, but the vegetable will remain succulent if planted near horseradish and other hot plants. Potatoes like marigolds, broadbeans, nasturtiums, cabbages, sweet corn and peas.

Strawberries like borage, lettuce and spinach. Beetroot likes onions and dwarf beans. Lupins and foxgloves planted all over the garden are very beneficial.

Beans do not like onion or beetroot, but they care for cabbage, cauliflower, celery, leeks, sweet corn and carrots. Yarrow and marigold are loved by all the garden. Cornflowers and violets do well together. Plant rosemary to keep off the carrot-fly and marjoram to drive off ants.

Roses love garlic, so plant a clove beside each bush. They will chase off every greenfly, and if they are prevented from flowering, they will help to enhance the perfume of the rose blooms. In drought weather, the garlic needs to be watered so that the rose roots will be able to continue to benefit from their sheltering presence. Sage, hyssop and mint keep caterpillars off vegetables, herbs and flowers. Nasturtiums will chase off woolly aphids and whitefly which infest fruit trees; let the tendrils wind up the trunks. Sawdust and oakleaves will drive away moisture-loving slugs – if they persist, a dish of beer is a kind way of eliminating them. Fern leaves scattered in the bed will deter strawberry weevil, and pine needles will improve the flavour of the strawberries. These fruits are unable to abide gladioli anywhere near them, however, and these flowers, even in the general vicinity, will hurt the flavour of the fruit.

❧

Elizabethan Herb Garden

You may like to lay out a little herb garden in the Elizabethan mode. This is simply done. Dig a round flowerbed, and build around it a little retaining wall of small bricks, placed in eight continuous sections to form an octagonal shape. This is not essential, but it raises the bed and separates it from the rest of the garden. Now

you must divide the bed into eight equal parts, separating them with conch shells or more bricks. In the middle of the bed, dig out a well in a ring two feet in diameter, and encircle it with bricks or shells. Put in sand, and plant rosemary there. The brick or shell 'spokes' of the octagon segments should extend from the lip of the rose-mary ring to the outermost edge of the herb-garden bed. Plant sage, mint, yarrow and parsley, borage, thyme, marjoram and dill in the segments, but not rue for this kills other flowers, particularly basil. Let rue stand alone in a shady nook somewhere in your garden where other flowers grow sparsely. Never plant the same kind of herb in the same spot as a habit, but change their positions in the segments, and replace a 'hot' with a 'cool' herb

81

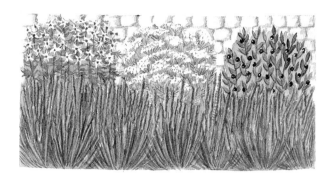

every season, for this encourages hardy growth and a rich, sweet aroma. If you make a magical herb garden in addition to your Elizabethan bed, make it bigger than the first, with lavender or lovage as a hedge, and vervain and other healers. If you grow henbane, hemlock and nightshade, all of which have magical and dangerous properties, stake them out well and name them with tags, for you must make no mistake in your dealings with them.

❧

TABLE OF HERBS ENDUED WITH MAGICAL PROPERTIES, AND OF OCCULT SIGNIFICANCE

Herb of Grace Rue
Herb of the Virgin Yarrow
Herb Trinity The Pansy (Heartsease)
Herb Sacra Vervain
Pantagruelion Hemp
Enchanter's Herb Nightshade
Sorcerer's Herb Henbane
Witch's Herb Monkshood (wolf's-bane)
Druid Herb Mistletoe

BASKET OF ROSES

Cut a flowerbed with softly scalloped edges and convex ends, and plant within it a host of miniature roses. Enclose the bed within sections of strong wire shaped to imitate a basket, or thin strips of wood in an openwork pattern with arches around the top to strengthen the impression of basketwork. Make a curving handle of wire which extends in a loop from front to back, and let this be entwined with a climbing rose. When the roses flower, the blooms will seem to be contained within a pretty basket. The rose-basket can be constructed around larger blooms, but would need to be more adventurous in scope.

DOROTHEA, PATRONESS OF GARDENERS

Gardeners have their own saint, but few know of her. She is Dorothea, a Christian martyr. Legend says that Theophilus, the secretary of the judge who condemned her to death, addressed her scornfully as she was going to execution, saying, 'Send me some fruit and roses, Dorothea, when you get to paradise'. Immediately after her death, while Theophilus dined with his companions, a young angel brought to him a basket of apples and roses, saying, 'From Dorothea, in Paradise,' then vanished. Theophilus was immediately converted.

St Dorothea is represented with a rose branch in her hand, a wreath of roses on her head, and roses with fruit by her side; sometimes an angel carries a basket filled with three apples and three roses. The rose basket in the garden was created as an idea in her honour, and all gardeners will benefit from remembering Dorothea, and calling on her, and her angel, who is the angel of the garden, for inspiration and aid. To dedicate a rosery to her and her angel is fitting, and can easily be constructed. Make a rockery, which will be the centre, and cover it with creeping roses. If there are fruit trees nearby, especially apple trees, let ramblers climb into them. Plant a circular bed, as large as your garden will allow, around the rockery, and create little brick paths in it, so that you may walk around the rosery for your delight and refreshment. Grow as many fragrant roses as you may in the bed, and let it be a place of sanctuary and repose. Do not forget to plant garlic in among your rosebushes, and lay under them copious amounts of fat and banana skins, burying these deep, for they will make your bushes flower in sweet profusion.

DOROTHEA'S LANGUAGE OF FLOWERS AND FRUITS

In honour of Dorothea, certain meanings have been given to roses and fruits. Here is her angelic basket of symbols:

Basket of roses 'Take the treasures of my heart'
Rosebuds 'Sweet thoughts of you shall dwell for ever in my bosom'
Garland of roses Lifelong blessings
Rose The loving heart of humanity
(*Austrian*) 'Thou art all that is lovely'
(*Bridal*) Happy love
(*Burgundy*) Pensive beauty
(*Musk*) Glad grace
(*Japan*) Beauty
(*China*) Everlasting loveliness
(*Damask*) Maidenly blushes
(*Moss*) 'I love you'

Apple Wealth and fecundity
Apricot Virgin bride and an early marriage
Cherry Storm-tossed love
Fig Inheritance
Grape Dreams and visions; worldly success
Lemon A broken bethrothal
Orange Danger
Peach Reciprocal love, good health, many blessings
Pear A new friendship
Plum A friend needs you
Raspberry Consolation
Strawberry A romantic rendezvous
Whortleberry Deception

THE MOSS HOUSE

In garden nooks, a moss house can be built. This must be made in miniature, of half a dozen rustic poles to support a sloping roof. The structure is best crude and rough. Knock laths close together between the poles, and cover all with an assortment of moss, which you must have, ready collected, from walls and forest floor, or wherever you can find it. Push the moss in between the laths (roots first) until it wedges firmly. Thatch the roof, and fix a circle of pine-cones as a cornice. Water well, and you will have a strange little house where perhaps your toad will sit, if you leave an opening for him. Fairy houses such as these can be an unusual and quaint feature of your garden, together with the garden grotto. This is an old idea, and is made by hollowing out a little cave in the garden, overgrown with everlasting pea, honeysuckle or ivy, and studded with shells and pebbles of all kinds, fronted by fern fronds. These little features are to please children, in the main, but they can beautify a garden, and give it a touch of eccentric mystery. If you build a grotto in your garden, keep a little lantern alight in it sometimes, well secured against fire, for the sake of Dorothea and her angel.

$\mathcal{I}NDEX$